Liveries of the WR Diesel Hydraulics

AF087774

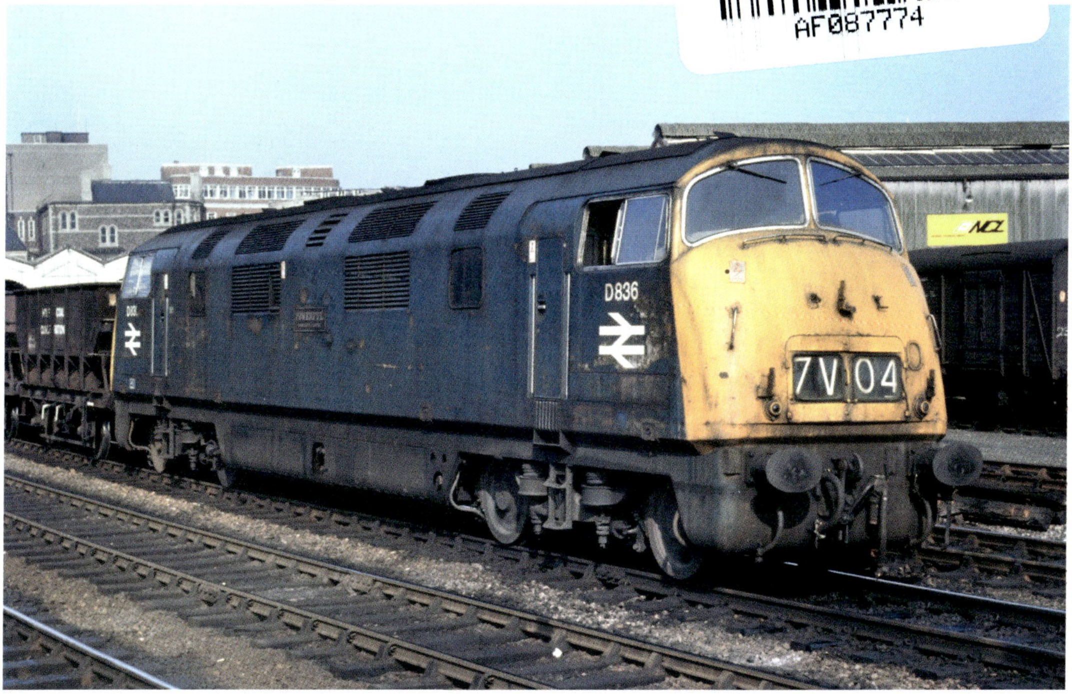

RUSSELL SAXTON

KEY Books

BRITAIN'S RAILWAYS SERIES, VOLUME 21

Published by Key Books
An imprint of Key Publishing Ltd
PO Box 100
Stamford
Lincs PE19 1XQ

www.keypublishing.com

The right of Russell Saxton to be identified as the author of this book has been asserted in accordance with the Copyright, Designs and Patents Act 1988 Sections 77 and 78.

Copyright © Russell Saxton, 2021

ISBN 978 1 80282 038 6

All rights reserved. Reproduction in whole or in part in any form whatsoever or by any means is strictly prohibited without the prior permission of the Publisher.

Typeset by SJmagic DESIGN SERVICES, India.

Contents

Introduction 4

Chapter 1 D600 'Warships' 5

Chapter 2 D800 'Warships' 10

Chapter 3 The 'Westerns' 36

Chapter 4 Class 22 62

Chapter 5 The 'Hymeks' 75

Chapter 6 Class 14 91

Introduction

The diesel hydraulic era on the Western Region of British Railways lasted less than 20 years, from 1958 to 1977, but it probably generated more interest in modern traction than anything else during and after the decline of steam traction. It is perhaps fair to say that the hydraulic classes went a long way towards reigniting interest in the railways for many who gave up when steam died.

One of the most interesting aspects of the hydraulic classes for me, and I suspect for many, was the bewildering variety of liveries they appeared in over such a short period of time, covering just about every possible variation of the old British Railways and corporate British Rail schemes at one time or another. Locos appeared in green, maroon, blue, and a few experimental liveries with and without yellow warning panels or full yellow ends, and in a wide range of styles with, of course, many individual differences.

At this point, a few space-saving acronyms used throughout the book might need to be explained as follows:

GNYP	Green No Yellow Panels
GSYP	Green Small Yellow Panels
GFYE	Green Full Yellow Ends
MNYP	Maroon No Yellow Panels
MSYP	Maroon Small Yellow Panels
MFYE	Maroon Full Yellow Ends
BSYP	Blue Small Yellow Panels
BFYE	Blue Full Yellow Ends

Some locomotives wore only one livery style during their careers, yet others carried many. In this book, I intend to list and illustrate every major and, hopefully, all minor variations of the liveries worn by this comparatively small percentage of the BR diesel fleet. There were many detail differences within the classes, but these will only be covered where relevant to the livery schemes. I have dealt with each class individually and the pictures are largely in chronological order for each type.

For me, being born too far north and too late in the era, I did not get to see much of the hydraulics in action, but my fascination with them was deep as I observed the changes from afar, via magazines and books. 'Westerns', 'Warships', 'Hymeks', D600s and North British Type 2s captivated my youthful imagination, and the changeover from the old green and maroon to the corporate Rail Blue liveries from 1966 onwards has remained almost an obsession with me, as anyone who knows me will testify.

I hope you will enjoy this look back over a fascinating era, both for the livery buff and those who just like some good old rail nostalgia. A special thank you is due to Ron Halestrap, Neil Phillips, and Martin Street for going the extra mile with this project and for the time and patience offered.

Chapter 1

D600 'Warships'

Deliveries of main line diesel-hydraulic classes began in January 1958 with D600 *Active*, the first of the five North British (NBL)-built A1A-A1A original 'Warship' design. These were followed from August by the smaller B-B version, numbered in the D800 series, later to become Class 42 if built by Swindon and Class 43 if built by NBL. They followed the then standard livery for main line diesel classes of Brunswick Green with a British Railways lion and wheel logo and serif numbers, prefixed with a 'D' to differentiate them from the then still-ruling steam locomotives, many of which carried the same numbers. D601 *Ark Royal*, in its first livery of GNYP, runs along the sea wall at Teignmouth on 2 September 1958, with the down 'Cornish Riviera'. The man on the sea wall seems not to be too interested in what, in 1958, would have been a rare sight – a main line diesel in action. (Rail Online/Rail Archive Stephenson/KL Cook)

D603 *Conquest* rolls into Truro on 15 May 1959, with the up 'Cornish Riviera', surprisingly minus the headboard. The grey roof employed on the D600s is seen to advantage in this shot. A GNYP diesel on a full chocolate and cream rake of coaches keeps the GWR atmosphere well and truly alive. This locomotive was also (I think uniquely) captured on film near Taplow, working a main line express in the 1961 film *Murder She Said*, albeit in black and white. Before long, it would be demoted from long-distance passenger work and confined for most of the rest of its life to the far west of the region. (Colin Moss Collection/Michael Mensing)

From the beginning of 1962, BR ordered that all green locomotives should have yellow warning panels applied to the front ends to aid visibility for permanent way staff. D604 *Cossack* was the first of the five NBL 'Warships' to appear as GSYP and is seen here ex works at Swindon in February 1962, with the panels freshly applied. D602/3 followed in April 1962 and then D600 in May. The last to receive them, D601 *Ark Royal*, actually ran in traffic for longer in GNYP than it did in any other livery, five and a half years from March 1958 to October 1963. (Paul Fuller)

The five NBL 'Warships' had headcode boxes fitted between 1964 and 1967. D600 had them applied during the same overhaul it went blue, being released from works in May 1967, and did not run as GSYP with headcodes like the other four locos. The old style of yellow panel with the raised centre was replaced by flat topped ones that came up to the headcode boxes. D603 *Conquest*, seen here at Plymouth on 21 August 1965, was a notable exception, as it had a larger than usual panel with the old raised centre that overlapped the headcode boxes. (Rail Photoprints/Richard Lewis)

From 9 June 1966, BR headquarters in Derby had sent out the instruction to repaint all locomotives into Rail Blue. Only a few hydraulics had been completed when Western Region permanent way staff complained that the full yellow ends applied to blue locos were less visible than the yellow panels. While this was investigated, Swindon began turning out blue locomotives with SYP from late-1966 onwards, and several 'Westerns', 'Warships', 'Hymeks' and D6300s were given the treatment. One of the original five D600s was in for a general overhaul in December of that year and was included in the scheme, D602 *Bulldog* being recorded at Margam during a short-lived spell in South Wales in August 1967. The bodyside arrow is a non-standard smaller version applied to many early Rail Blue hydraulics. (RCTS Image Archive/Michael Mensing Collection)

All five of the original NBL D600 'Warships' are lined up on the scrap line at Laira on 23 April 1968. D604 *Cossack* is at the front, followed by D601 in GSYP, D602 in BSYP, D603 with its larger yellow panel and, at the rear, the only BFYE example, D600. The TOPS classifications were introduced by BR in the summer of 1968, and while these engines are often referred to as Class 41, they were withdrawn some months before the scheme was made public and never known as such during their lifetimes; any reference to such being on paper only. (KDH Archive/Keith Holt)

Once the dispute over visibility was resolved, one more of the class, D600 *Active*, received a coat of Rail Blue, complete with the by now standard full yellow ends, at a general overhaul in May 1967, only to be withdrawn in December, along with rest of the class. Despite the short period of time it ran in traffic, a mere seven months, and at the height of the frenzy to photograph the last steam workings hundreds of miles north, it was still fairly well recorded. Here, D600 rolls into Saltash in 1967's 'summer of love', with an unidentified passenger working. (Barry Knapp)

D600 *Active* again, seen after withdrawal alongside its classmate D601 *Ark Royal* (which retained its GSYP livery, as did D603/4), dumped at Dai Woodham's scrapyard at Barry in February 1970, shortly before it was cut up. There was little interest in preservation for diesels in 1970, and it is not surprising that four of the five D600 'Warships' were cut up fairly quickly after withdrawal. However, D601 somehow lasted for another decade amongst the rows of steam engines at Barry before tragically being scrapped in 1980, well into the era of diesel preservation. The other three were scrapped at Cashmores, Gwent, in 1968. (Rail Photoprints)

Chapter 2

D800 'Warships'

Swindon 'Warship' D801 *Vanguard* is seen at Old Oak Common in October 1959 in as-delivered condition, namely Brunswick Green with the numbers on the cabsides in serif style and the BR logo centrally placed above the nameplate, which were red-backed on a green locomotive. The first few into service, D800–12, had discs rather than the four-character headcodes. Yellow warning panels were some way off in the future and all of the 'Warships', except the last few to be delivered, NBL-built D859–65 from January 1962 onwards, entered service without them. (Grahame Wareham)

From D813 onwards, the class was fitted from new with four-character headcodes. D869 *Zest* is seen passing Churston, on the Kingswear branch, with the 09.05 from Liverpool on 26 July 1962. The engine is still in as-delivered condition minus yellow panels, although 'Warships' began to have them applied from January 1962. However, it took until September 1965 for the entire fleet to receive them. (John Whiteley)

The first livery modification to a 'Warship' was to NBL-built D845 *Sprightly* in August 1961. A small yellow surround was added to the headcode panels and a white 'eyebrow' above the windscreens, as seen here at Starcross on 18 August 1963. This version of the warning panel was not adopted, and the more familiar larger style was used instead. *Sprightly* later went into standard green with a normal size yellow panel in February 1964, and eventually into Rail Blue in February 1970. (Author's collection)

The first 13 'Warships' that were built without four-character headcode panels were all retro-fitted with them between September 1963 and September 1965. All except D803 *Albion* had small yellow panels applied beforehand, as shown here with disc-fitted D807 *Caradoc* arriving at Paddington on 30 July 1964. D803 had headcode panels and SYP applied at the same time in February 1965 and was the sole example of the D800–12 batch not to run in traffic with yellow panel and discs. (Rail Photoprints/Chris Davies)

The classic lines of a green-liveried SYP Swindon 'Warship' are displayed by D826 *Jupiter* at Waterloo on 19 September 1964, with an Exeter working. As late as this, a few of the first batch (D803/7/8/10) were still running with discs instead of the four-character headcodes and, In the case of D803, minus the yellow panels too, as was headcode-fitted D856. (David Christie)

D855 *Triumph*, of the NBL batch, is seen in GSYP at Exeter St Davids on 18 September 1964. D855 was to go maroon with panels in May 1966, and then to Rail Blue in January 1970. (Brian Ireland)

The nameplate and British Railways' lion and wheel of D851 *Temeraire* at Plymouth in May 1967. The lion and wheel faces left, as it did on all the GSYP 'Warships' except one, this being D811, which had a right-facing emblem on one side only for an unknown reason. (Colin Moss)

For a short while from May to September 1965, all 71 'Warships', of both varieties, theoretically wore the same livery of BR green with panels. Whether all 71 were in traffic at once is debatable, however. D833 *Panther* of the NBL batch, later to be designated Class 43, is seen on 12 September 1964, near Reading West with an unidentified working. *Panther* would remain in green with panels until September 1969. (David Christie)

During the brief period when all 'Warships' were GSYP, D831 *Monarch* runs along the sea wall at Dawlish in early August 1965. D856 was released from Swindon into traffic on 18 May 1965 as GSYP, and D857 was taken into Swindon on 31 May, having failed a few days beforehand. It was released in maroon in September, so it was presumably in works in GSYP at the time of this shot. The period for all 71 potentially being in traffic as GSYP was a week or so at most! (Dave Flitcroft)

On the same day as the shot of D831, D866 *Zebra* rounds the curve at Dawlish, with a westbound freight. *Zebra* would remain GSYP until September 1968 and did not appear in maroon. (Dave Flitcroft)

As with the 'Westerns' discussed later, the Western Region (WR) decided to repaint 'Warships' into maroon from September 1965. The whole fleet may have been done eventually, but the introduction of Rail Blue in 1966 halted the scheme after 32 had been repainted, D801/2/5/6/9/11–3/15/17/21/23/28/29/32/34/38–40/42/44/48/55/57/58/61–3/65/67/69/70 being the ones concerned. The lion and wheel was replaced by a coaching stock roundel in the same place and the white bodyside stripe also vanished, giving the locos a slab-sided appearance. Just over a year after the application of maroon, and still looking quite smart, D801 *Vanguard* pauses at Basingstoke, with a Waterloo–Exeter working on 10 May 1967. Sadly, it had not long to go before withdrawal in August 1968. Of the first three pilot scheme 'Warships', D800 was withdrawn while green, D801 maroon and D802 Rail Blue. (Rail Online)

Although it looked very smart when freshly applied, the effects of the WR washing plants and, no doubt, sea air along the coast of Devon and Cornwall took a toll and some of the maroon locos looked very weather-beaten after just a few months. A spectacularly battered example, D862 *Viking*, is seen at Bristol Bath Road in October 1969. Repainted in maroon in November 1965, it would not to last much longer in this colour, receiving full Rail Blue in April 1970. (John Wiltshire)

As only 32 of the 71 'Warships' went into maroon before Rail Blue repaints began in late 1966, many carried on in green for some time. D814 *Dragon*, in GSYP, passes Deepcut in Surrey, with the 07.20 Exeter–Waterloo on 15 April 1967. *Dragon* had only a short time left in this livery, as it was repainted into the first standard style of Rail Blue in June and would later wear the final style from May 1969. For the time being though, it could be enjoyed in its green livery on this wonderfully liveried rake of assorted Southern and BR coaches in green and blue/grey. (David Rostance)

This is the only known photo of D848 *Sultan* in green without the horizontal stripe and with a coaching stock roundel, as a maroon loco would have worn. Seen here at Dawlish on 5 March 1966, the loco's livery is something of a mystery. It was done like this sometime after August 1965, most likely as a depot patch-up to cover some cosmetic damage, and it is not even known if both sides were the same. Before long, the loco would go to Swindon and emerge in June 1966 carrying maroon with small yellow panels, a livery it wore until the end. (Geoff Landon)

The only 'Warship' to be withdrawn in green livery, and one of only two not to have the words 'Warship Class' on its nameplate, was D800 *Sir Brian Robertson*, seen here at Basingstoke in 1967. Delivered as GNYP in August 1958, it had yellow panels applied in May 1963 and then the four-character headcodes in March 1964. Withdrawn in October 1968, it was another locomotive that ran as GNYP without headcodes longer than it ran in any other livery. (Rail Online)

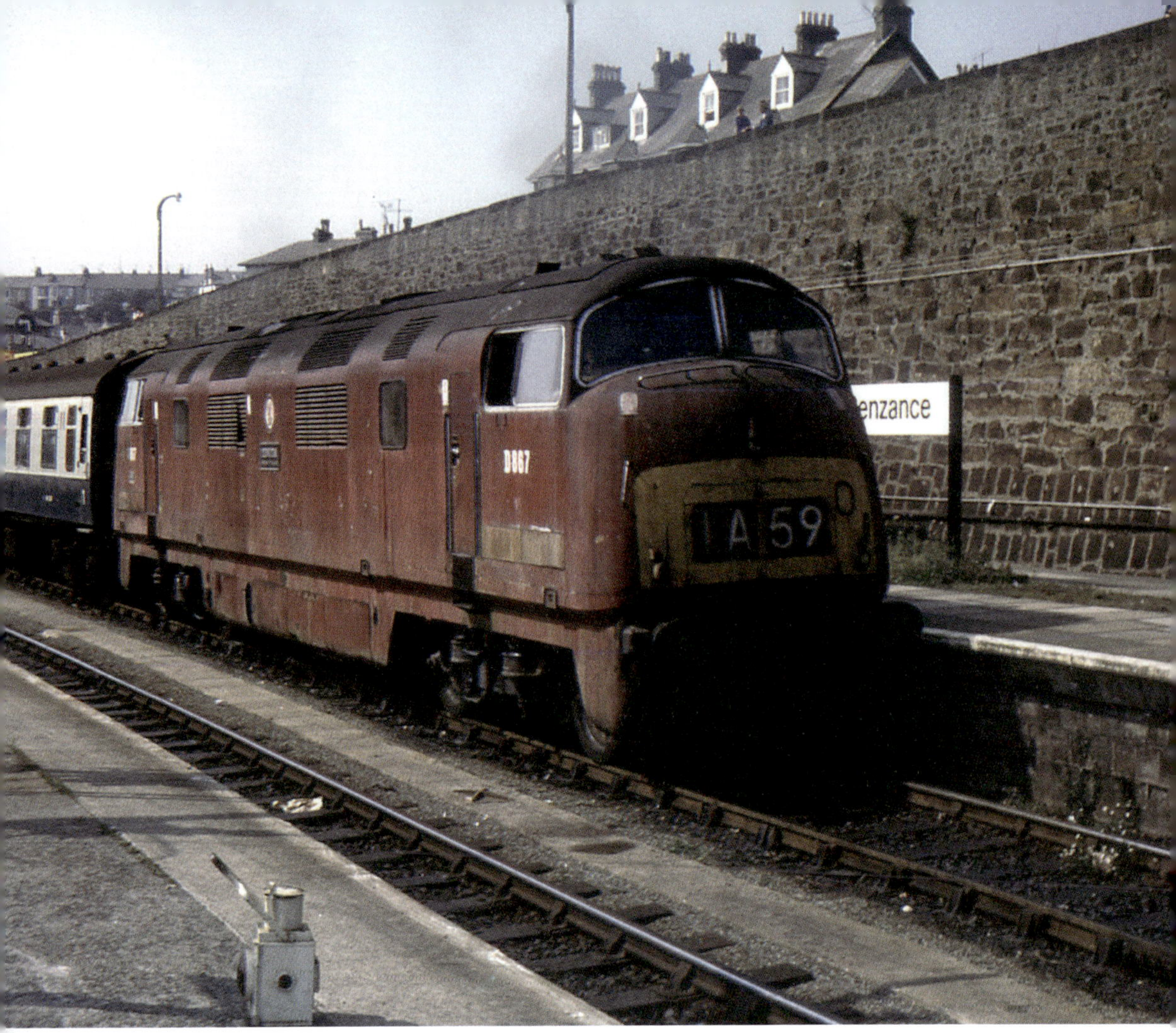
enzance
D867
1A59

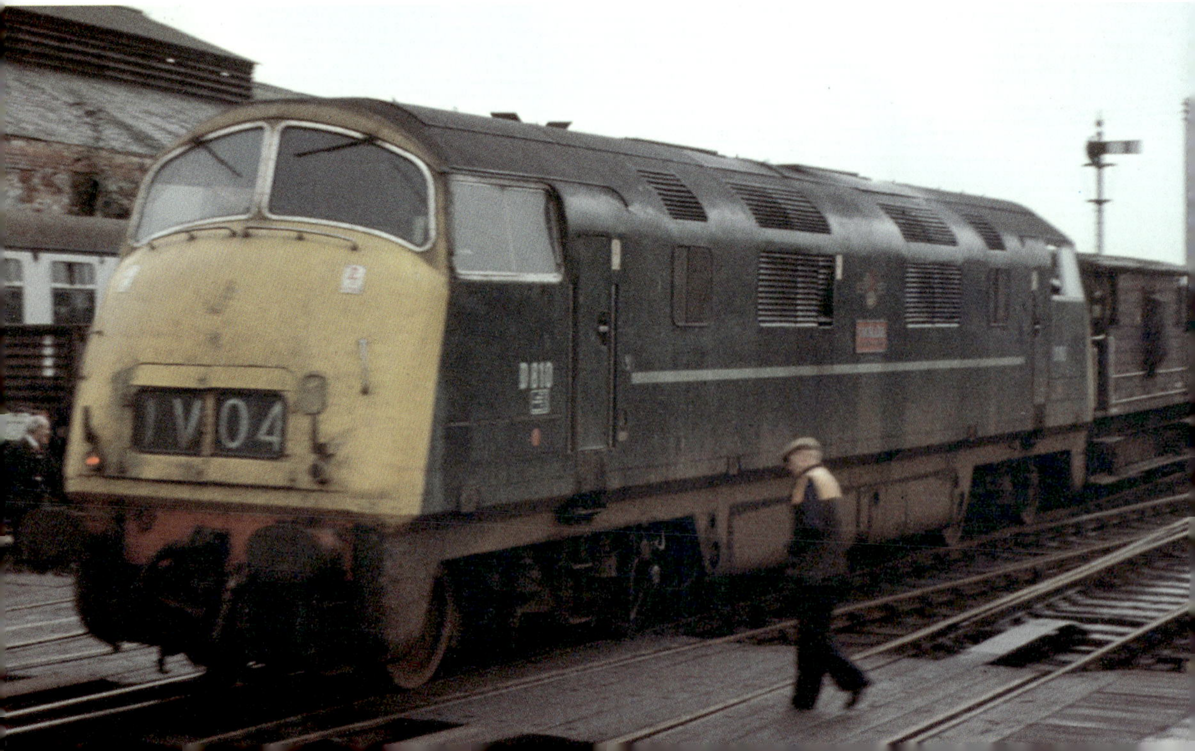
1V04
D810

D800 'Warships'

Opposite above: Almost the last 'Warship' to run in maroon with small yellow panels was D867 *Zenith*, which is seen at Penzance on 29 August 1970, awaiting departure with the 12.25 to Paddington. There were still a few maroon 'Warships' around with FYE for some time to come, but this loco went into Rail Blue in October 1970, leaving D861 *Vigilant* to soldier on in traffic until January 1971, when it went into Swindon. Emerging in Rail Blue in March, D861 was withdrawn that October, and became the 'Warship' that ran the shortest time in that livery. D801, D840 and D848 never gained FYE on their maroon livery and were scrapped in MSYP, albeit before this date. (Andy Kirkham)

Opposite below: Only two 'Warships', D808 and D810, received GFYE in December 1967 and January 1968, respectively. D808 *Centaur* only ran until July 1968 before going blue, but D810 *Cockade* lasted in the style until May 1970, becoming the last green 'Warship' in traffic. Photos of D808 are scarce indeed and D810 is not much better represented, but here it is at Newton Abbot in early 1970. Classified 42 (Swindon) and 43 (NBL) under the BR scheme introduced in 1968, 'Warships' had a small data panel applied on the driver's cabsides; this listed class, weight and route restriction, amongst other data. This was first applied to D838 *Rapid* in August that year. (Chris England)

Below: From February 1967, it was ordered that all main line locomotives were to have full yellow ends applied, although this took many years to carry out after an initial flurry of activity and some never did. The first maroon 'Warship' to get FYE was D817 *Foxhound* in August 1967, shown here arriving at Exeter St Davids on 22 May 1970, from the LSWR route, presumably from Waterloo. D817 never lost the livery and was withdrawn as MFYE in October and cut up the following March. (Fred Castor)

Liveries of the WR Diesel Hydraulics

Opposite above: Many other maroon 'Warships' had full yellow ends applied, including D809 *Champion*, seen at Newton Abbot in June 1969. Note the data panel, which was always blue, regardless of the livery of the locomotive it was applied to. MFYE locos were D805/6/9/11/12/15/17/23/29/32/34/38/42/44/69/70. 'Western' D1055 behind has already had the D prefix painted out, which became common on the class, along with the 'Hymeks', following the end of steam in 1968. (Chris England)

Opposite below: D809 *Champion* is seen again, this time at Gaer Junction, Newport, with a freight on 27 May 1971. The locomotive never lost its MFYE livery, and a few others also went to the breakers in this style. As well as D809, D815, 817, and 838 remained MFYE to the end. Sixty-three of the 71 'Warships' went blue; the eight remaining were D800 in GSYP, the four MFYE examples, and the three MSYP. (Steve Cresswell)

Below: D815 *Druid* looking weather beaten in MFYE at Exeter St Davids on 27 July 1971, especially when compared to the neat all blue/grey rake of coaching stock. Coaches went into the corporate colours far more quickly than locomotives, and the sight of maroon 'Warships' and 'Westerns' on all blue/grey rakes was a common one as early as 1967. *Druid* was given maroon with yellow panels in October 1966, and then full yellow ends in January 1968, remaining in MFYE until withdrawal in October 1971. The loco thus wore four different liveries in its career: GNYP, GSYP, MSYP and MFYE. (Jim Sparks)

The very first 'Warship' into Rail Blue, after the instruction went out to repaint the whole fleet, was D864 *Zambezi* in November 1966. Initially, it had incorrect non-corporate serif numbers on the bodysides, instead of on the cabs, as it should have done, along with Burnt Umber underframes. Accident damage in early 1967 necessitated repairs and it appeared in the first standard version of Rail Blue, with cabside serif numbers above the arrow. It did, however, retain the non-standard umber valancing. (RW Carroll Collection)

After D864 appeared in BFYE with the bodyside numbers in serif style, the next blue 'Warship' was D831 *Monarch*, this being repainted at the end of November 1966, shortly after D864. As stated in the caption to D602, WR permanent way men had complained that the full yellow ends were less visible than the small panels, and Swindon then began to outshop the various hydraulic classes in Rail Blue livery but with yellow panels. D831 was the first of two 'Warships' to fall under the umbrella of the scheme, and again incorrect serif-style numbers were placed on the bodysides, rather than the cabs. The cabside arrows were of the correct size, though. The loco is seen at Waterloo on 3 June 1967, in BSYP, looking somewhat anachronistic working alongside the last few 'West Country' and 'Merchant Navy' steam locomotives. (Neil Avent Collection)

The other BSYP 'Warship' completed in December 1966, D830 *Majestic* was a slightly closer attempt at the specified standard Rail Blue with cabside (correct) serif numbers (incorrect), small yellow panels (incorrect), and a smaller than normal arrow above the nameplates in the centre of the bodysides (also incorrect). Far more camera shy than D831, due to the short time it spent in this livery, just December 1966–August 1967, shots of it are rare and working ones doubly so. Here, D830 is seen at Cullompton in the spring of 1967, with the 13.35 Paignton–Sheffield. (Sid Staddon Collection/Amyas Crump)

The next two 'Warship' repaints after D830/1 were D846/7, done in January and February 1967, respectively. Both had FYE as was correct and the numbers in the right place on the cabsides as D830, albeit again the non-corporate serif ones, but the smaller incorrect size arrow was used above the nameplates again. D846 *Steadfast* stands at Old Oak Common in company with 'Hymek' D7058, which is also of interest, as it has the early style of FYE with the yellow not wrapping round onto the front as far as later repaints. The shot is undated, but as both engines have data panels, it is sometime after September 1968, and prior to D846's repaint into standard blue in June 1970. (Fred Castor)

The companion loco to D846, D847 *Strongbow*, rounds the curve at Hereford on 20 April 1969, while working the LCGB 'Woodpecker' railtour from Waterloo to Bulmer's of Hereford, and taking in a few other places en route. I suppose *Strongbow* was the only possible choice! A minor detail difference between this and D846 was that the numbers were closer to the cab doors on D847, as can be seen here. *Strongbow* ran from February 1967 to withdrawal in March 1971 without any subsequent repaints, but D846 was given the final version of Rail Blue, with central bodyside arrow and a 'D'-less Rail Alphabet number on the cabsides, in June 1970. (George Woods)

A close-up of the nameplate of D847 *Strongbow* showing the undersized double arrow symbol; this was taken on the same LCGB 'Woodpecker' tour as illustrated in the previous caption, in April 1969. It is possible that the smaller arrows were hand-painted, rather than decals. (George Woods)

Two more 'Warships', D819 *Goliath* of the Swindon batch and D857 *Undaunted* of the NBL lot, were also non-standard in yet another variation. Repainted blue in April 1967, they had FYE and cabside serif numbers as per the previous few, but with the incorrect smaller style BR double arrows on the cabs. D819 is seen at Woking on 15 June 1967, with a Waterloo–Salisbury service. (HN Forsyth)

Above: Eventually, Swindon got it (almost) right, regarding the instructions for Rail Blue and the standard style of 'D'-prefixed serif numbers with the larger arrow on the cabsides, often referred to as BFYE1. Strictly speaking, the serif numbers were incorrect and only Swindon and Inverurie were applying them to blue locomotives throughout 1967. D859 *Vanquisher* was the first into the style in June 1967, followed quickly by the aforementioned D864 *Zambezi*, after its short-lived spell in the first incarnation of blue, although it retained the umber underframes. The next was D818 *Glory*, also in June 1967, and complete with the specified blue underframes, as seen at Truro in June 1970. D802-4/7/13/14/16/18/20/27/41/49/53/59/63/64/68 all ran in this livery before the style of lettering changed to Rail Alphabet in January 1968. D818 later morphed into the final Rail Blue style and acquired red-backed nameplates in the summer of 1972, before gaining much notoriety, as will be discussed later. (Michael Taylor)

Opposite above: Easily the least photographed 'Warship' in Rail Blue was D863 *Warrior*. It was not the shortest-lived in blue livery, that title goes to D861 *Vigilant*, which only ran for a matter of months in 1971, but as D863 ran from December 1967 to withdrawal in March 1969, its tenure in blue coincided with the end of steam. As a result, it was rarely photographed in blue and hardly ever while working. Here it is at Oxford in September 1968, with the 15.15 Paddington-Hereford. (Doug Nicholls)

Opposite below: The style of Rail Blue applied to 'Warships' varied from year to year. From January 1968, 'Warships', of both the Swindon and NBL varieties, were outshopped with the Rail Alphabet style of numbers instead of the serif, and cabside large size arrows, known as BFYE2. Although no steam locomotives had run for over a year with three- or four-digit numbers, the 'D' prefix lasted right until the end of steam traction. D808/19/20-2/26/35-7/43/49-51/60/66 appeared in this variant between January and October 1968. Here is D836 *Powerful* on a coal train at Newport on 13 April 1971. (Steve Cresswell)

D800 'Warships'

A contrast in Rail Blue styles on the dump at Swindon Works on 14 August 1971. The first blue 'Warship' D864 *Zambesi*, in BFYE1 with the serif style numbers, faces D860 *Victorious* of the NBL batch, in 'BFYE2' with the later Rail Alphabet style numbering. Both had been withdrawn in March 1971, and would be scrapped in November (D864) and December (D860) of the same year. (Rail Online)

The last three 'Warships' to appear in BFYE2 were not repainted until late in 1968, just after BR had discontinued the use of the 'D' prefix, and thus emerged as plain 803, 825 and 856. None of the three got a subsequent repaint and this was the final livery for them. 825 *Intrepid* is seen at Exeter St Davids on 5 June 1971. (Derek Jones)

Many 'Warships' ran in green, maroon and blue in their careers. D826 *Jupiter* escaped the maroon stage, going directly to blue in January 1968, and was the first to appear in the Rail Alphabet style. However, by 5 May 1971, as seen at Exeter, its former green livery was showing through its very worn coat of blue. It was not long before it got a Laira makeover to make it look slightly more presentable. The depot painted out the cabside arrow to fall into line with the final style, but reverted to the serif numbers, albeit minus the 'D'. The final result is depicted with the shot of 868 a little later on. (Derek Jones)

Another well-worn example was D824 *Highflyer*, whose black-backed nameplate is wearing away to reveal the previous red colour from its time in BR green, as it waits at Paignton, with the 15.55 to Paddington on 6 November 1972. (KDH Archive/Keith Holt)

D822 *Hercules* is also showing some wear as it arrives at Exeter St Davids on 12 September 1970, with the 13.10 from Waterloo. Its former green livery, lost in March 1968, has had two years of sea air and WR washing plants to allow it to show through, as does the former red nameplate. (Derek Jones)

The final style of Rail Blue applied to the 'Warships' was a central arrow on the bodysides and cabside Rail Alphabet numbers, minus the 'D', of course. This is known as BFYE3 and was applied to all Class 42 and 43 'Warships' outshopped from February 1969 onwards. 805 *Benbow*, seen at Newport on 23 June 1972, illustrates the style, which was also worn by 806-8/10–4/16/18/20/21/23/24/27–9/31–5/39/41–6/52–5/57/58/61/62/65/67/69/70. Some of these were on their second repaint and had previously run in the earlier blue styles. (Jim Sparks)

D800 'Warships'

Many 'Warships' ran in more than one variety of Rail Blue, but only D820 *Grenville* wore all three standard versions: BFYE1 from August 1967, BFYE2 from May 1968 and finally BFYE3 from September 1971, being the last to get this livery. It ran for a little over a year in BFYE3, as it was withdrawn in November 1972. D820 is seen at Laira on 22 May 1970, during its BFYE2 period. (Fred Castor)

823 *Hermes* poses at Exeter St Davids on 1 August 1971 in BFYE3, which was acquired in September 1969. Previously, *Hermes* was a MFYE loco and also appeared in GNYP, GSYP and MSYP, this being five liveries in a little over nine years. Note the naval plate next to the cabside number, which was unique on a 'Warship'. (Jim Ireland)

810 *Cockade* is also in its final style of Rail Blue at Exeter in August 1972. A mere four liveries were carried by this one: GNYP, GSYP, GFYE and BFYE3. *Cockade* was one the last 'Warships' in traffic, lasting until December 1972. (Jim Ireland)

The other Laira patch-up, 868 *Zephyr*, was done in the style of D826 *Jupiter*. *Zephyr* is shown here as it leaves Exeter St Davids on 31 July 1971, with the 15.55 Paignton–Paddington. Like *Jupiter*, it has the cabside serif number, minus the 'D' prefix, and the arrow re-positioned from the cab to the bodyside. The nature of the patch-up is evidenced by the position of the data panel to the side of the site of the former cabside arrow. (Derek Jones)

829 *Magpie* gained a small level of celebrity due to it often featuring on the ITV children's show of the same name, which was a rival to BBC's *Blue Peter*, which, of course, had its namesake in A2 Pacific. By July 1972, 829 *Magpie* had run in BFYE3 for almost three years, and had already been withdrawn in January that year, only to be reinstated in March, but minus its nameplates. It is seen in this condition at Barnstaple on 28 July 1972. There was much made of the loco, or at least its nameplates, being acquired by the TV programme but, sadly, nothing came of the plans, and it was scrapped at Swindon in January 1974. (Dave Thorpe)

Above: The 'Warship' that ran the least time in Rail Blue was D861 Vigilant, which was repainted from maroon to Rail Blue as the last example to be so treated and emerged from Swindon in March 1971. Pictured here at Exeter St Davids on 22 May 1971, it has only five more months to go before withdrawal in October. (Derek Jones)

Opposite above: The last 'Warships' rode the rails in December 1972, but scrapping was not immediate for some, and a few lingered at Swindon Works for a few years afterwards. D832 *Onslaught*, the last 'Warship' to be withdrawn, outlasted them all in BR service, as it was acquired by the Research Centre at Derby and used for some years as a dead weight in test trains. Long after the end of the class in normal service, *Onslaught* stands at Derby on 26 May 1976, being snapped from a passing train. (David Hayes)

Opposite below: D818 *Glory* at Swindon, long after the hydraulic era ended, on 6 June 1981. This locomotive did the opposite to the rest by going from Rail Blue back to green. It was withdrawn in November 1972, near the end of the type's reign, but it was not cut-up; instead, it was used as a source of spares for D832 and was adopted by Swindon Works as a 'pet'. It was initially repainted in shiny Rail Blue in the BFYE3 style in September 1975, for the works' open day on the 13th of that month, and then restored to GSYP livery in October 1980 and used as a demonstration loco for apprentices. Tragically, after the works was given its closure notice in 1985, the loco was scrapped, some say as an act of spite by angry works staff. (Russell Saxton)

D800 'Warships'

Chapter 3
The 'Westerns'

Above: The first 'Western', D1000 *Western Enterprise*, was originally delivered in an experimental livery of Desert Sand in December 1961. The colour was originally suggested for the 'Deltics' by BTC designer George Williams but was applied to a 'Western' instead. The experiment was not a success it appears, as it remained the sole example to run in this livery, and the same can be said of the other experimental livery of Golden Ochre applied to D1015 and featured later. At first, the engine had no yellow panels and red bufferbeams, plus the window pillars were done in black. It also had cast aluminium crests, as applied to AC electrics on the secondman's cabsides, where the rest of the fleet had the roundel, and it retained this after receiving standard MSYP until its repaint into Rail Blue in June 1967. Here it is at Shrewsbury in May 1962. (Rail Photoprints/Hugh Ballantyne Collection)

Opposite above: D1000 *Western Enterprise* had the addition of yellow panels on its Desert Sand livery in February 1963, and it ran like this until October 1964, when it appeared in standard MSYP. Here it is at Old Oak Common in April 1964, looking well-worn in the company of green 'Warship' D867 *Zenith*. D867 was also destined to go MSYP a couple of years down the line in September 1966, one of the last few to do so. (Grahame Wareham)

Opposite below: Several 'Westerns' were delivered in 1962 in maroon minus the yellow panel, this being D1001/5–10 of the Swindon build, and D1039–43 of the Crewe batch. They had curious yellow patches applied to the bufferbeams, which differed on the Swindon and Crewe varieties. The Swindon ones were all as per the pictured D1005, while D1039/40 of the Crewe batch had the yellow extended to cover the lip around the recess, whereas D1041–3 just had it extended to the edge of the same. This was a short-lived variation. D1005 *Western Venturer*, seen here at Swindon on 12 August 1962, ran for only six months like this, from introduction in June to January 1963, and none ran past mid-1963, when the last received the classic 'Western' livery of MSYP. One other minor variation was D1001, which had white window pillars, while all other maroon locomotives had off white/grey. (Rail Photoprints)

The 'Westerns'

Above: One of the Crewe-built 'Westerns', D1041 *Western Prince*, speeds past Solihull during its short spell in MNYP livery from October 1962 to May 1963. Subsequently, D1041 would run in MSYP and then MFYE from May 1968, before receiving the inevitable Rail Blue FYE from October 1969. The other MNYP 'Westerns' lost this interesting livery variation as follows: D1001 in October 1962, D1005 in January 1963, D1006 in December 1962, D1007 in November 1962, D1008 in January 1963, D1009 in July 1963, D1039 in May 1963, D1040 in June 1963, D1042 in May 1963, and D1043 in November 1962. (RCTS Image Archive/Michael Mensing Collection)

Opposite above: Seven 'Westerns' were delivered in green livery, all with SYP from the off; D1002–4 of the Swindon batch and the first few into service of the Crewe build, D1035–8, but it was a fairly short-lived affair for all concerned. D1002, seen here at Paddington on 30 July 1964, was delivered in March 1962, and ran a mere three years before receiving MSYP in May 1965. D1003 lasted until November 1965 before going maroon, but the other Swindon example, D1004, lasted until February 1967 and went directly to Rail Blue with FYE, being one of only three never to run in maroon. Green 'Westerns' had red bufferbeams and red-backed number and nameplates, unlike the black-backed ones on the maroon examples, which perhaps improved the look somewhat! I never thought green looked right on a 'Western'. (Rail Online)

Opposite below: The four Crewe green-liveried 'Westerns' went onto a variety of liveries; D1035 to MSYP in February 1966, and D1038 the same in January that year, but D1036 and D1037 never ran in maroon, and went to blue with small yellow panels in November 1966 and January 1967, respectively. D1036 only ran for one year as BSYP and went BFYE in November 1967, but D1037, shown here while still green at Burnham on 17 September 1966, lasted until June 1971 as BSYP before the inevitable application of blue with full yellow ends. None of the green locos ever got full yellow ends, which would have been interesting! (Rail Online/Dave Cobbe)

The 'Westerns'

D1036 *Western Emperor* was always easily recognisable by its higher-than-normal cabside numberplate, this being due to the fitting of AWS equipment inside the cab. It did not have long to go in GSYP in this view at Newton Abbot on 18 September 1966, just a few more months before it became one of the seven BSYP locomotives. Of equal interest, especially at this late date, is the Class 22 beside it. The loco is one of just three delivered new with headcodes in GNYP (D6334–6) and is most probably D6334, which ran as GNYP until late in 1966, and quite possibly into early 1967, when it went into Swindon, later to emerge in Rail Blue in the spring. It is highly likely it never ran as a GSYP machine. D6335/6 did run as GSYP, although I have no exact dates. Disc headcode D6333 was another, minus yellow panels, into 1967 and was also probably never GSYP. (RCTS Image Archive/Mike Burnett Collection)

The shortest-lived livery variation on a 'Western' was on the other experimentally liveried locomotive, Golden Ochre D1015 *Western Champion.* D1015 had a 'T' bar yellow panel applied to its 'A' end, which was only worn for four days from 21–25 January 1963, while the engine was on display for the press, as seen here at Swindon Works in the famously snowy winter of that time. The Swindon Works manager had the choice of which 'Western' would be painted in this livery, the other loco in the frame was D1014 *Western Leviathan*. (Grahame Wareham)

D1015's 'T' bar panel was replaced with a non-standard yellow panel (incidentally, it had one the same at the 'B' end) for its release to traffic after the display photos, and it ran in this livery until a repaint into MSYP in November 1965. Here, it arrives at Reading on 5 August 1964 with an unidentified working. (Rail Photoprints/Chris Davies)

From August 1962, it was decided that all subsequent 'Westerns' would be painted in maroon livery with black bufferbeams and roofs, with the exception of one, which was to be given the Golden Ochre livery. The maroon is the classic livery of the 'Westerns' and was worn by all but three of the class. It is wonderfully illustrated by Swindon-built D1026 *Western Centurion*, photographed as it heads for Paddington at Severn Tunnel Junction on 9 September 1965. D1026 wore just two liveries, MSYP, as seen here, from introduction in 1963, and Rail Blue with full yellow ends from June 1967 to the end. Note the chocolate and cream stock in the consist, which was becoming a rare sight by 1965. The whole GWR atmosphere remains very much in place, which I imagine was the WR's intention all along. (Bill Wright)

Maroon locomotives had the British Railways roundels on the secondman's cabsides, as seen here on Crewe-built D1068 *Western Reliance*, awaiting departure from Paddington on 21 June 1964. The sole exception was D1000 *Western Enterprise*, which had an aluminium crest on the cabsides, as per the AC electric locos on the London Midland Region, which it wore up until its repaint into Rail Blue in June 1967. (John Ireland)

Another classic MSYP 'Western' shot with D1069 *Western Vanguard* at Paddington on 30 April 1966. At this time, most of the fleet were in this livery, for which they are perhaps best known, with just D1004/36/37 bucking the trend and remaining in GSYP. In a few months, the first of the class would begin to appear in blue livery with SYP or FYE, depending on the loco involved, and many maroon locos would receive full yellow ends. For a short time, it was possible to see 'Westerns' in maroon and green SYP and blue with both full yellow ends and panels. (John Ireland)

D1070 *Western Gauntlet* leaving Paignton on 11 July 1968, with a working to Bradford, looks oddly out of place on a rake of blue/grey-liveried Mk.1 stock. Coaching stock went blue/grey at a much faster rate than the locomotives did, and this was a very common sight, even in 1968. D1070 went to BFYE in February 1969 and wore just the two liveries during its career. (Antony Guppy Collection)

By 1969, the 'Westerns' could be seen in maroon or Rail Blue, both with and without panels, while all the green examples had been repainted by 1967. The sight of a MSYP example at Birmingham New Street was not a common one, and this is the only photograph I have seen of one there. D1042 *Western Princess* backs onto the stock of the 17.25 Manchester Piccadilly–Cardiff, which it will have taken over from an AC electric. *Princess* wore just two liveries, this and full Rail Blue FYE from September 1970 until it became one of the first few to be withdrawn in 1973. (Antony Guppy Collection)

Western Region locos were often dirtier than the ones on other regions, and D1007 *Western Talisman* shows how dirty a maroon 'Western' could get if allowed as it leaves Cardiff Central on 22 April 1967, with a service for Plymouth. Note the clean patch at the bottom of the cabside, this showing the true colour underneath! D1007 was to remain in maroon until October 1970 and only ran in Rail Blue for a short time, as it was involved in the Ealing accident in late 1973 and withdrawn as a result in January 1974. (Robert Masterman)

Full yellow ends became mandatory on all locomotives from February 1967, and while none of the green Class 52s ever wore them, several of the maroon ones did, as seen here with D1001 *Western Pathfinder*, at Plymouth on 14 March 1970. Full yellow ends were applied to this loco in May 1968 and it ran in this condition until November 1970, when it went to Rail Blue FYE. (Chris England)

The first 'Western' to get FYE was D1056 *Western Sultan*, shown here at Exeter St Davids in 1968. Not only was this engine the first to appear in MFYE in September 1967, it was also the last to remain in the livery, not going Rail Blue until April 1971. MFYE 'Westerns' were D1001/2/8/12/16/25/39/41/44/45/54/56/67/68. (Fred Castor)

A great shot from the camera of Derek Jones showing a MFYE 'Western' and 'Warship' together, showcasing examples of the only two diesel classes to wear this pleasing livery combination. D1016 *Western Gladiator* stands next to D842 *Royal Oak* at Gloucester on 8 September 1969, nearing the end of its tenure as MFYE from March 1968 to July 1970. The 'Warship' spent a similar time in the style, from May 1968 to October 1970. (Derek Jones)

The first 'Western' into blue was D1030 *Western Musketeer*, released from works on 2 August 1966, and seen here at Taplow goods shed open day on 17 September that year, in the company of green 'Warship' D837 *Ramillies*. Although the main attraction was undoubtedly the ex-GWR steam locomotives, no less than 7,000 people attended this event! Swindon erroneously repainted D1030 with small size arrows (believed to have been hand painted, no less), red bufferbeams and small yellow panels, much to the annoyance of BR headquarters at Derby. (Antony Guppy Collection)

A close-up of the cab of D1030 at Taplow, again showing the undersized double arrows and red bufferbeams. D1030 was spray painted, which was an unpopular choice with Swindon's painters, apparently, and subsequent repaints were done the traditional way with a brush. The spray painting did appear to wear rather better than a brush coat and, three years later, D1030 looked in better condition than most of the other early blue locomotives. (Grahame Wareham)

The second 'Western' into blue in November 1966 was D1048 *Western Lady*. This one was done in the correct style, with full yellow ends, black bufferbeams and a large sized double arrow. Here it is a few years later, on 31 May 1975 at Teignmouth. (Trevor Ermel)

After D1048 appeared in BFYE, Swindon then turned out a few locos in BSYP in late 1966–early 1967, owing to the aforementioned dispute over PW visibility. Recently ex works D1017 *Western Warrior* is seen at Didcot on 12 February 1967. Interestingly, the loco appears to lack the red route restriction discs that it had later in this livery style. D1017/36/37/43/47/57, as well as the aforementioned D1030, ran in this style, often dubbed 'Chromatic' blue. However, it is in fact exactly the same shade of blue as standard Rail Blue, and the illusion of a lighter shade is mainly caused by the SYP and the variations of 1960s film emulsions. In the case of D1030, the spray finish also appears slightly lighter in the daylight. (Bryan Hicks)

In the summer of 1968, BR began to use the TOPS classification Class 52 for 'Westerns' and applied the data panels, which were always in blue, regardless of the colour of the locomotive it was given to. D1063 was the first 'Western' to receive one in October. BSYP D1047 *Western Lord* is seen at Exeter St Davids in October 1969, with a data panel applied below the number (note it is the same shade of blue as the body), alongside the red circle denoting the route restriction. (Chris England)

D1047 *Western Lord* is seen again at St Budeaux in June 1969 in BSYP. The seven BSYP locos received BFYE as follows: D1017 in March 1968, D1030 in April 1970, D1036 in October 1967, D1037 in June 1971, D1043 in December 1968, D1047 in July 1971, and, last in both numerical and chronological order, D1057 in January 1972. After D1057's repaint, the entire fleet of 74 locomotives were all in standard Rail Blue. (Chris England)

The shortest-lived 'Western' in BSYP was D1036 *Western Emperor*, which only ran a little under one year as such from November 1966 to October 1967. As a result, it was often missed off lists of the BSYP locos for many years after due to the brief time it spent in the style. The higher placed cabside numberplates remained so throughout the life of *Emperor* and its three liveries of GSYP, BSYP and BFYE. D1036 ticks over at Old Oak Common on 12 February 1967, as, down the road at Swindon, classmate D1004 had just emerged as the second 'Western' in BFYE. (Rail Online)

After the PW visibility dispute was resolved, Swindon reverted to outshopping blue 'Westerns' in BFYE as per D1048, and the next out was the formerly green-liveried D1004 *Western Crusader*, done in February 1967, a week or so after D1017 was released in BSYP. By early January 1972, all 74 'Westerns' were in Rail Blue FYE, as shown on D1051 *Western Ambassador* at Newport on 7 June 1972. This instance of the entire class intact in one livery was a short-lived thing, as, in a little under a year, withdrawal of the Class 52s would begin and the hydraulic era would be almost over. D1051 only wore two liveries in its career, both very much standard styles; MSYP from 1963–67 and BFYE from September 1967 to the end of its service in September 1976. (Jim Sparks)

Rail Blue D1063 *Western Monitor* poses at Laira on 19 August 1973. By now, withdrawals of the 'Westerns' had commenced, although D1063 had three more years to run in traffic. New in maroon with small yellow panels, D1063 changed colour just once in October 1968 to BFYE. It had one minor claim to fame, as mentioned earlier in the book, in that it was the first 'Western' to receive the BR data panel on the driver's cabsides. (Jim Ireland)

D1037 *Western Empress*, another of the trio that never ran in maroon livery, pauses at Oxford in June 1974. By the standards of the date, the loco is in very good condition. The 'D' prefix has been completely removed from the cabside numberplate, which happened to a number of the class. *Empress* wore three liveries in its career: maroon with panels, blue with panels, and the final Rail Blue from June 1971. (Grahame Wareham)

D1036 *Western Emperor* **retained the distinctive higher position for its numberplate into BSYP and, later, BFYE liveries, right up to the end of its career in November 1976. Some of the 'Westerns' were given another coat of blue by Laira after Swindon ceased overhauling them, and on the day of its release following a repaint after a 'B' exam, 29 May 1975, D1036 stands at Exeter St Davids in absolutely immaculate condition, ready to depart with an unidentified working. You could probably still smell the paint! (Trevor Ermel)**

D1036 had worn a bit by the time it was photographed at Plymouth on 18 October 1975, but it retains its 'D' prefix, as several did to the end. (George Woods)

D1046 *Western Marquis* rolls into Paignton on 30 May 1975. This was the last maroon 'Western', remaining in MSYP until May 1971, by which time the rest of the fleet was Rail Blue, with D1037/47/57 retaining the small yellow panels for a few more weeks in the case of D1037/47, and a few months in the case of D1057. (Trevor Ermel)

D1043 *Western Duke* rounds the curve at Teignmouth on 31 May 1975, with the 'Cornish Riviera Limited'. Another former BSYP loco, *Duke*, had been in Rail Blue since December 1968 and had run in three liveries, MSYP, BSYP and BFYE. (Trevor Ermel)

Above: A few 'Westerns' had small experimental air vents fitted to the front ends, locos were D1012 (3/71), D1028 (4/71), D1039 (2/71), D1056 (4/71) and D1071 (2/71). The vents were square, except on D1039, which were rectangular. Here is D1028 *Western Hussar* on a parcel working at Reading on 1 August 1973. The vents were fitted in April 1971, but this locomotive had been Rail Blue since July 1969, and ran minus the vents for nearly two years. D1071 had been blue since 1967 without them, but the rest were given them at the same time as blue livery. (Andy Kirkham)

Opposite above: Minor detail differences on the cabs of D1023 *Western Fusilier* and D1037 *Western Empress* at Old Oak Common. D1037's shed sticker is a modern (post-autumn 1973) 'LA' one applied above the number, but D1023's is the older style 'Laira' one and applied to the bodyside, as was done up to that time. (Antony Guppy)

Opposite below: The usual state of a 'Western', especially in the last year or two, is shown by D1057 *Western Chieftain* at Laira on 13 December 1975. The loco still has the WR red route restriction disc on the cab, a rarity by 1975, although some were replaced by depot staff. As mentioned earlier, D1057 was the last 'Western' to receive BFYE livery in January 1972. Previously, it was one of the BSYP group and actually ran as BSYP longer than it did in either maroon with panels or in BFYE! (Steve Jackson)

The 'Westerns'

D1058 *Western Nobleman* stands at Old Oak Common on 22 February 1976, the sole hydraulic round the famous turntable on this day. From 1 January 1976, BR had decided that the headcode display was redundant and all locos had them wound back to read '0000'. D1058 was a fairly early BFYE repaint in November 1967 and ran for over nine years in the style until withdrawal in January 1977. (Antony Guppy)

D1041 *Western Prince* stands at Westbury on 17 April 1976. Its tatty external condition is not really a surprise as, after a repaint into BFYE in October 1969, it was to receive no further attention, and its next visit to Swindon was after withdrawal in February 1977. Happily, it still exists today in somewhat better condition. (Stephen Burdett)

D1048 *Western Lady* passes Solihull on 17 April 1976, with the 12.25 Birmingham–Paddington, complete with the then near obligatory window hangers, enjoying the last few months of Class 52 haulage. Although they were supposed to read '0000', many of the class had the headcode blinds wound to show the loco number, as seen here, making them easily recognisable from some distance. Besides being the first 'Western' into BFYE in December 1966, D1048 also ran in the livery for the longest spell, clocking in over ten years. (David Rostance)

Another Class 52 with the square air vent, D1056 *Western Sultan* is seen at Bristol Temple Meads on 20 June 1976. It is also another loco with the headcode panel wound to display the loco number, as became common after the headcodes were officially abandoned from the start of 1976. (Tom Harper)

Above: As it was often used on railtours towards the end of its career, Laira spruced up D1013 *Western Ranger* with white wheel rims, silver bufferheads, headboard clips and screw couplings, plus red-backed name and numberplates. It retained these trimmings until the end of its career in 1977, but did not always work the top jobs, as shown here on a humble freight working at Reading on 10 August 1976. (Geoff Dowling)

Opposite above: D1013 *Western Ranger* is seen again on a more prestigious working as it rolls into Exeter St Davids on 6 July 1976, with a service from Paddington to the west of England. This loco had been BFYE since August 1968, one of the longest to run in this livery. (Steve Jackson)

Opposite below: The Western Region's washing plants were often the cause of the poor state of many hydraulic liveries. However, whatever is to be said about washing plants at loco depots and the effect on paintwork, this clean-up of 1023 *Western Fusilier* at Laira has to be more of a tender loving care job. The occasion was the 'Western Sunset Special' railtour on 9 October 1976 from Swansea to Plymouth, and 1023 had gone to Laira for a wash and brush-up, as well as a refuel. (Steve Jackson)

The 'Westerns'

D1013 again, easily identified by its red name and numberplates, is seen at Penzance on 13 November 1976, on the return 'Cornishman' railtour from Derby to Penzance and back. This was 'Peak' hauled from Derby to Birmingham. (Steve Jackson)

The numberplate of D1046 *Western Marquis*, seen here dumped at Laira on 13 December 1975, just two days after withdrawal, has been removed for safekeeping and to be sold by BR. Where is it now, I wonder? This reveals something that has puzzled many; the former maroon livery is shown, but the '6' is a serif style painted on character. Crewe used to apply serif style numbers without 'D' prefixes to the cabsides, while the locos were in primer and the numberplates were attached later. Swindon went one better (of course) and used the same font as the cast numbers, with the 'D' and all lined up with chalk marks! (Steve Jackson)

Many BR locos had their headcode panels replaced with the 'domino' panels after January 1976 (and in the case of the 'Deltics', much sooner), but the sole Class 52 to get them was D1023 *Western Fusilier*. This is sometimes reported as having been done to comply with Eastern Region regulations and to allow the loco to work a railtour there in November 1976, but this is apparently not the case and was just due to the poor condition of the existing headcode blinds. *Fusilier* is seen passing Old Oak Common on 5 February 1977, with the return leg of the Monmouthshire Railway Society's 'Capitals United Express' from Cardiff to Paddington and back. (David Rostance)

The last hurrah for the diesel hydraulic era came on 26 February 1977. D1010 *Western Campaigner* and D1048 *Western Lady* run off Bristol Bath Road depot in readiness to follow the 'Western Tribute' railtour on its way from Swansea to Plymouth. These locomotives were standing by in case either D1013 or D1023 were to fail while hauling the special train. D1010 has had the 'D' prefix removed from the numberplate, but D1048 retains it. The new order stands on the left! (Thomas Harper)

Chapter 4

Class 22

The North British Type 2s, later to be classified as Class 22 in 1968, were all delivered in green livery with grey roofs and red bufferbeams, D6300–36 as GNYP and D6337–57 as GSYP. D6300–33 had disc headcodes, as typified by this undated view of D6326 at Swindon Works. D6334–6 were delivered new, minus yellow panels but with headcodes. The GNYP locos had the yellow panels applied from 1962 onwards, but a few ran into the late-1960s without them, and it is possible one or two never received them at all and went straight to blue. As well as GSYP being applied to the non-yellow panel locos, the four-character headcode boxes were fitted to all except D6301, which was withdrawn as GSYP in December 1967, with the discs still in place. (Rail Online)

D6327 is still to receive its yellow panels as it arrives at Plymouth, assisting Castle 5057 *Lord Walgrave* on a working from Paddington on 1 June 1963. (KDH Archive/Keith Holt)

GNYP D6324 rolls into St Agnes with a freight from Newquay on 15 August 1962, complete with headcode boxes, which were fitted after an accident, while running in tandem with D6302 in January 1961. Both locos were sent back to NBL for repairs and returned with headcode boxes fitted, these being of a different type to those fitted from new, not being flush with the cab front and sited higher up than normal. D6302 had yellow panels applied in August 1962, with D6324 following sometime later; both had oversized versions that came up to the headcode boxes. (Rail Online)

D6318 stands at Newton Abbot on 18 September 1966 in GSYP and, despite the late date, retains the disc headcodes. This locomotive was repainted in BFYE during 1967, and I do not know if it had the headcode boxes fitted at the same overhaul and ever ran in GSYP with them. There were several locos still minus the headcode boxes at this late date, D6301/28/32/33, at least, and possibly others. (RCTS Image Archive/Mike Burnett Collection)

A few other Class 22s ran with the early style of headcode boxes after 1965 besides D6302/24, namely D6306/7/17/26, as seen here on D6317 at St Blazey in June 1966, but these engines either had the yellow panels beforehand or were given them simultaneously with the headcode boxes. D6302/24/26 were later modified with the flush style headcodes, but D6306/7/17 were scrapped with the non-flush style. (Brian Ireland)

A contrast in front end styles with standard GSYP D6330, on the right, next to D6317, on the left, with the older style of headcode boxes. Note the different positioning of the overhead warning flashes due to the differing front end layouts. D6317 was withdrawn in this livery, but D6330 went onto become Rail Blue in 1970. (Fred Castor)

Above: D6330 again in the summer of 1970 at Exeter St Davids in the company of 810 *Cockade*, recently repainted in blue and until recently the last 'Warship' to run in green. There was a short period to capture these two together in these liveries, as 810 emerged in blue in May 1970, and D6330 in December, probably going into works in late October. (Fred Castor)

Opposite above: A very interesting and unusual shot of D6323 at Seaton Junction, working a milk train on 27 March 1966. At an unknown date prior to September 1964, D6323 was fitted with a spare cab in the style of the D6334–57 batch, but the place where the headcodes would be are just blanked off, and the discs/gangway doors are retained. As far as is known, this was unique. (Tony Martens)

Opposite below: More than half the Class 22s were withdrawn in green livery, all but two in GSYP, and all but one with headcodes. D6308 rests at Newton Abbot in October 1969, looking withdrawn. However, this was not the case, and, in fact, it was to receive a general overhaul in 1970 and emerge in Rail Blue. Data panels have been applied to the driver's cabside while the yellow WR route availability circle remains on the secondman's side. The worksplates have vanished, however! (Chris England)

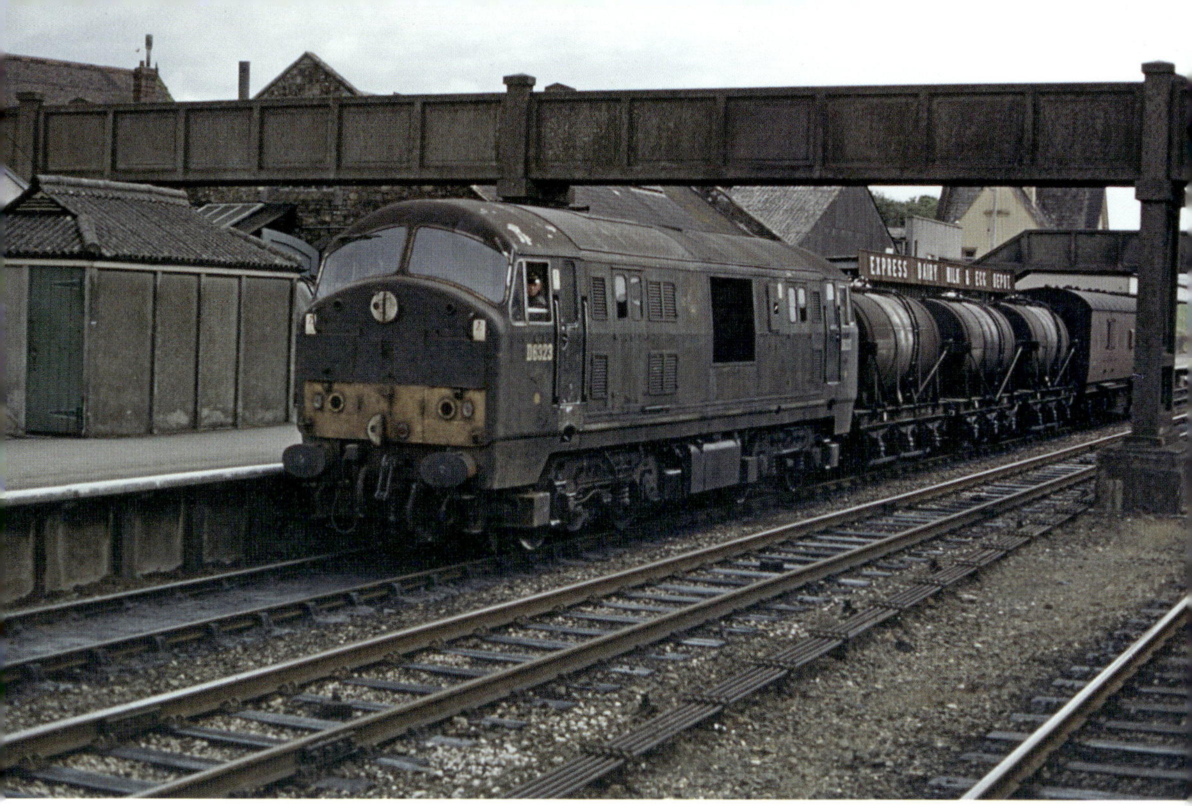
EXPRESS DAIRY MILK & EGG DEPOT
D6323

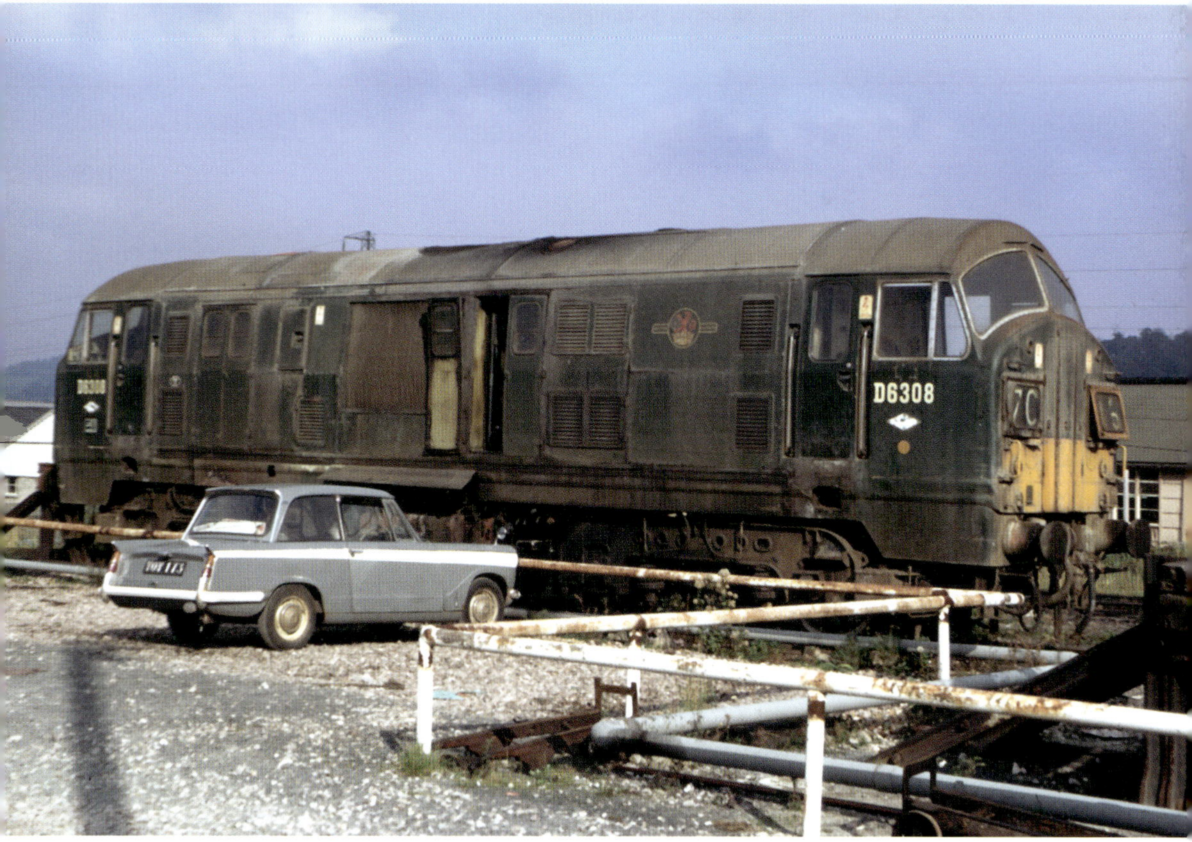
D6308

Four Class 22s were repainted in Rail Blue with small yellow panels during the time of the PW visibility dispute in late 1966–early 1967: D6300/3/14/27. D6303 of the first six pilot scheme locos is seen at Laira in June 1967, not too long after overhaul and still quite shiny. On the two pilot scheme BSYP repaints, the arrows were in the same position each side, but on D6314/27 they were in different places, owing to the different body design. On one side, the arrows were to the left of the large radiator grille but, on the other side, they were positioned inboard of the secondman's cab door, after the small grille. (Gordon Edgar/Charlie Cross Collection)

D6300/3 were withdrawn in May 1968, fairly soon after repainting, and D6314 in April 1969, but D6327 soldiered on in the livery until withdrawal in May 1971. It is already looking careworn at Paddington in this view on 7 April 1969, and it continued to deteriorate. It has recently had the data panel added to the cabside and retains the WR yellow route availability disc on the other end. D6300/3 were withdrawn before there was time to do this. (Rail Online)

Once the dispute of PW visibility had been settled from early 1967, new blue repaints on Class 22s had cabside arrows and serif 'D' prefix numbers with full yellow ends. D6343, seen at Laira in July 1971, follows the style as applied to D6302/18/22/25/28/32–4/36/37/39/40/42/43/54. After these were done, overhauls ceased for a while, as withdrawals began from December 1967. (David Hancox)

BSYP D6327 is nose to nose with BFYE D6354 on the scrap line at Bristol Marsh Junction on 14 August 1971. Although the Rail Blue was applied just a few months apart, there are many differences; D6327 retains its BSYP, with cabside number in serif and a bodyside arrow symbol, whereas D6354 shows the second style of Rail Blue applied to the Class 22s, with the serif 'D' prefixed number above the arrow on the cab. To finally lay the myth of 'Chromatic Blue' to rest, it will be observed that both locomotives are exactly the same shade of blue! (Jim Sparks)

Usually, after the 'D' prefix was abandoned in September 1968, most diesel locos had them painted over or removed. This was obviously more difficult with the metal plates on the 'Westerns', and the cast numbers of the 'Hymeks', but the painted numbers on most other classes were fair game. The Class 22s were a notable exception to this rule. None of the 1966–67 blue repaints, some of which ran until 1971, lost the 'D' and only two of the green examples did, namely D6323 and D6352. 6323 sits on the scrap line at Swindon in April 1972, just under a year after withdrawal, and a few weeks before it was to be broken up. (Grahame Wareham)

The other green Class 22 to lose the 'D' prefix, 6352, was not scrapped in green but repainted Rail Blue in 1970, as part of the second batch to receive the livery. The locomotive stands at Gloucester Horton Road on 28 May 1969, looking fairly tatty, but not bad, by Class 22 standards in 1969. Whoever has painted out the 'D' has used a different shade of green, which was quite common. Some locos of other classes had it done in blue, but at least it retains the NBL diamond worksplate above the recently applied BR data panel for a while longer. D6306 had the 'D' painted out on the driver's cab on one side, but not anywhere else. Modellers beware! (Derek Jones)

Only two Class 22s ran with full yellow ends on green livery, D6312 and D6331. The latter is seen two months after withdrawal at Gloucester Horton Road on 22 May 1971. The yellow ends were applied to this loco around April 1970, most likely at depot level. You would expect the 'D' prefix to have been painted out at the same time, but this did not occur. (Derek Jones)

The other GFYE Class 22 had the yellow ends applied much sooner than D6331, as far back as December 1967 by Swindon Works, after some minor collision damage repairs to the 'B' end. Notice that the cabside numbers are a little higher than other green Class 22s, as the driver's card holder had been removed. Another minor variation was that it had the number of worksplates reduced from four to two, and they were sited on the central valances, like the BFYE examples. Here it is at Plymouth on 27 August 1970, in the company of Class 46 191. Both GFYE Class 22s were withdrawn in green in 1971. (Jim Sparks)

Shortage of motive power, due to the non-arrival of the promised diesel electric replacements, prompted the WR to put a few Class 22s through works in 1970 and early 1971. All the selected machines (6308/19/26/30/38/48/52/56) were green beforehand, none of those done in 1967 were overhauled again. The livery was slightly different, with no 'D', of course, while the Rail Alphabet number style was employed and the arrow/number position was reversed. Here is 6326 at Southall on 27 June 1970. (Gordon Edgar)

An example of the variety of liveries on Class 22s extant in 1970 is shown here at Gloucester Horton Road. D6354 is in the 1967-style BFYE, GSYP D6320 and D6310 follow, and, at the rear, GFYE D6331. D6354 attained minor fame in the early 1970s, as it was featured on the closing credits of the news and chat programme *ATV Today*, until someone pointed out it had been withdrawn, and it was replaced with footage of a steam locomotive, much to my youthful disappointment. Diesels were still unpopular with many enthusiasts at the time! (Derek Jones)

Repainted blue in July 1970, 6348 contrasts in style with D6337, done in late 1967, as both stand on the scrap line at Swindon in May 1972. It seemed odd that recently overhauled locos, such as 6348, had such a short active life after this outlay; it had just one year until it was withdrawn in July 1971. (Grahame Wareham)

6356, in the final style of Rail Blue applied in March 1970, stands at Exeter on 22 May 1971. Perhaps the most bizarre incident of a Rail Blue repaint was classmate 6319, which was given a full overhaul and released to traffic on 10 June 1971, and was then withdrawn in September, a mere three months later. As is well known, it was the subject of a preservation attempt, but was cut up by Swindon in November 1972 in error, by which time all of the rest of the class had been scrapped. A sad loss to preservation, especially as it had been so recently overhauled. (Derek Jones)

D6340, one of the 1967 repaints with serif numbers and 'D' prefix, stands on the scrap line at St Philip's Marsh on 10 January 1972. Still in good external condition considering it would not have had a repaint since 1967, it was somewhat of an exception to the rule; most of them were quite tatty by withdrawal. (Derek Jones)

Chapter 5

The 'Hymeks'

Almost brand new, 'Hymek' D7002 is seen at Salisbury on 29 August 1961, in the livery style of the first 20 members of the class. D7000–19 had no yellow warning panels and the attractive green livery with pale green solebars and white window surrounds. From D7020, 'Hymeks' were delivered with small yellow panels from new and, starting with D7018, the GNYP examples were retro-fitted with the panels. Eventually, all 101 appeared in this style and some never lost it, although other members of the class appeared in many of the six liveries possible for a 'Hymek'. (Nigel Kendall)

Above: D7005, delivered new in September 1961, was still minus the yellow warning panels as late as April 1964, when photographed at Nailsea with an unidentified working. The addition of panels was the sole change to the livery of this locomotive in 11 years of service. It is featured later in this book at the other end of its short career, just seven years after this shot, and still in GSYP. D7017 was the last 'Hymek' to run without yellow panels, they were applied a few weeks after this shot, circa July 1964. (Robert Carroll Collection)

Opposite above: Then almost brand new, GSYP 'Hymek' D7085 pauses at Cardiff, with the 12.00 Portsmouth–Swansea in July 1963. The locomotive is already starting to gather dirt after only one month in service, as WR locos seemed to do so more than any other. Red route availability stickers are on the cabsides at both ends. D7085 would go Rail Blue with full yellow ends in the spring of 1968, another that wore only two liveries throughout its life. (Chris Gwilliam)

Opposite below: The pristine state seen on D7085 was not to last long, and most of the 'Hymeks' soon looked very battered indeed. D7028, seen at Old Oak Common in April 1970, is not too bad by 'Hymek' standards, despite a few scratches on the cab corner. It has acquired the data panel on the driver's cabside but retains the red route availability disc on the secondman's end. Before much longer, D7028 would succumb to Rail Blue and run in traffic until 1975, as one of the last to do so. (Grahame Wareham)

The 'Hymeks'

Above: A few 'Hymeks' gained full yellow ends on their green livery from 1967, as shown by D7092 at Gloucester Horton Road on 30 September 1969. The cast alloy number has had the 'D' prefix painted over but not actually removed, which was quite common. Normally, on any 'Hymek' with FYE, the strip of aluminium acting as a kickplate for the full-width chequer plate step on top of the bufferbeam cowl was painted black, but this loco retains the unpainted version. The data panel is on the bodyside, which was also unusual. The GFYE members of the class were D7000/9/13/14/16/18/20/23/31/84/92–4/97. Most got repainted into Rail Blue, with the exception of D7013/14/20, which were scrapped in GFYE. D7075 is often included in published lists of GFYE locos, but I have not unearthed anything concrete to prove it in 40 years of searching, and I would stick my neck out and say it probably never was. (Derek Jones)

Opposite above: By June 1972, locomotives in GSYP of any class, hydraulic or otherwise, were conspicuous by their rarity. Just two locomotives remained in traffic in green with panels out of the entire WR hydraulic fleet at this time, both 'Hymeks', namely D7005/54. They were not the last locomotives on BR to run in the style, as there were still a few foreigners from other regions, including 'Peaks', Class 40s and Class 25s, along with assorted departmentals extant at this time and for a while to come. However, the days of green diesel hydraulics were almost over. D7005 had only a couple of weeks left to run when it was photographed at Newport on 29 June 1972, leaving D7054 to carry on as the last surviving green 'Hymek', and last green hydraulic, for another five months until December. (Jim Sparks)

Opposite below: A total of 13 'Hymeks' went to the breakers still in green livery: D7002/3/5/6/8/13/14/20/21/24/25/54/60. By 1972, interest in modern traction was beginning to grow and the last green survivor, D7054, became quite a celebrity, especially as it remained in its GSYP livery to the end of its life. Here it is a couple of years before it became an enthusiast favourite, at Oxford on a freight in November 1969. (Grahame Wareham)

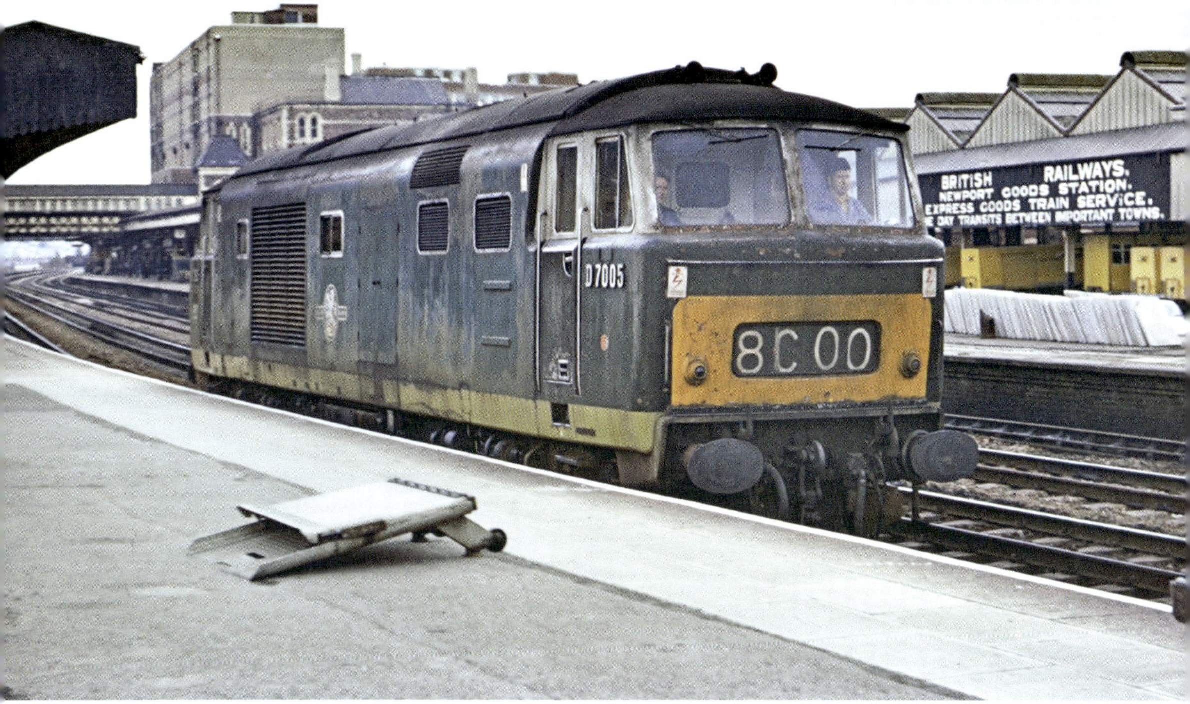
D7005
8C00
BRITISH RAILWAYS.
NEWPORT GOODS STATION.
EXPRESS GOODS TRAIN SERVICE.

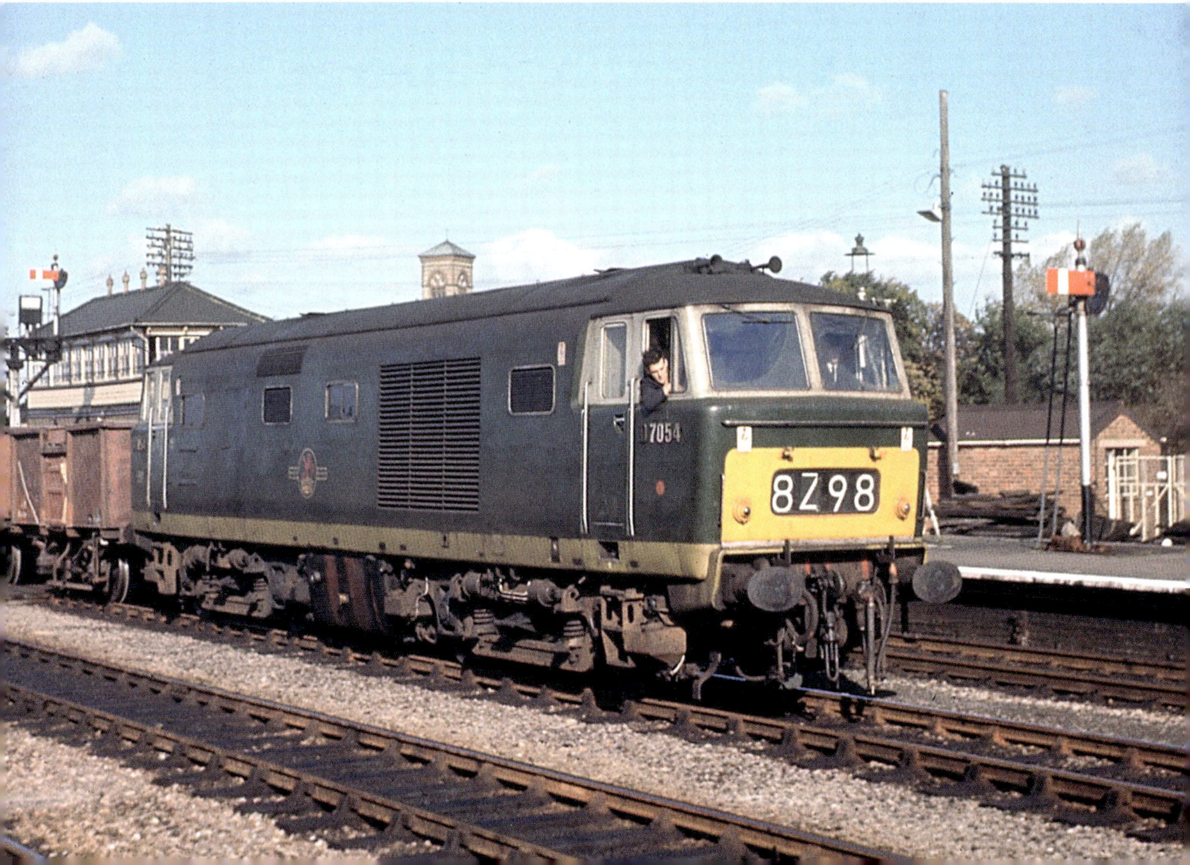
8Z98

D7013 lies derelict at Swindon Works in May 1972 in green with full yellow ends, one of three 'Hymeks' to be withdrawn in this livery style. Bar the SYP and then FYE, probably the sole addition to its as-delivered paint scheme was the application of the data panel (in blue) on the cabside. Withdrawn in January, it was to be scrapped in October 1972. Just edging into the shot is the first 'Hymek' to be repainted blue in 1966, D7033, displaying the shallower yellow ends applied to early repaints. (Grahame Wareham)

D7014, another of the trio to be withdrawn in green with full yellow ends, is seen dumped at Bristol Bath Road on 12 April 1972. It was soon to be moved to Swindon and broken up in August 1972. (Jim Sparks)

The first 'Hymek' into blue was D7033 in November 1966. The next few were in the BSYP style, due to the PW visibility dispute, but D7033 and the few that appeared in BFYE up to the summer of 1967, varied slightly from what became the standard Rail Blue pattern. The yellow ends did not wrap around the front so far as on later repaints and the bodyside arrow was sited a little higher up the sides on D7033/38/63, at least. Recently, ex works D7038 stands at Bristol Bath Road in July 1967. (David Bromley)

The non-standard yellow ends were applied to D7004/12/33/35/37/38/42/58/61–3/67, of which D7004/12 were formerly in BSYP, while the others went from green to BFYE directly. D7058 has the shallower FYE, but the arrow is in the normal place in this April 1969 shot at Old Oak Common. This suggests it straddled the change from the non-wrap around FYE to the full version and the re-positioning of the arrow lower on the bodysides; either that, or different painting teams at Swindon did things their own way! (Grahame Wareham)

The next three 'Hymeks' into blue after D7033 were D7007 and D7051, done in November 1966, and D7004, done in December. These three were, as shown here on D7007 at Worcester in April 1969, Rail Blue (not Chromatic!) with small yellow panels and no lining to the window surrounds. It is a matter of opinion if this style is considered attractive or not, but the loco behind has the later variation of BSYP, which for my money was an improvement. The bodyside arrow is sited higher up than later became standard, as with all the first few blue 'Hymeks'. (Rail Photoprints/Mike Jefferies)

D7007 and D7051 of the BSYP with blue window surrounds never received any update to the livery and were both withdrawn in the colour scheme in April and January 1972, respectively. D7007 is pictured almost at the very end of its career on 12 April 1972 at Bristol. Data panels have been applied to the driver's cabside and the WR red route classification disc has been patched over with a different shade of blue. It was to be withdrawn within days of this shot. (Jim Sparks)

It was soon decided to apply white window surrounds to the next run of BSYP 'Hymeks', and one of the three plain blue surround locomotives, D7004, was quickly given them as soon as March 1967, when the loco was photographed at Oxford. It was in full Rail Blue with FYE by July 1967! D7004 appeared in green without panels, green with yellow panels, blue with panels and plain windows, blue with panels and white window surrounds, and, finally, in BFYE, a record five different livery schemes for a 'Hymek'. (Grahame Wareham)

After D7004/7/51 were repainted BSYP, Swindon carried on overhauling and repainting 'Hymeks' into blue with panels for a few months. However, it was decided that the look of the locomotives were vastly improved by keeping the white window surrounds as per the green engines, and quite a few emerged from works in this quite pleasing style: namely D7010/2/27/34/36/40/46–8/52/56/57/59/64, plus, of course, D7004 that had formerly had plain blue window surrounds. D7027 passes through Oxford in this November 1969 view, showing the livery to fine effect. (Grahame Wareham)

The BSYP livery seemed to wear less well than the standard Rail Blue and most of the BSYP 'Hymeks' were in poor external condition by the end. D7004, D7012, D7027, D7048 and D7064 were given FYE (D7012 had the non-wrap around variety), but the rest were withdrawn as BSYP. D7004/48 also gained the normally placed arrow, while the others retained the higher sited arrow when in standard Rail Blue. D7048 had to have a full repaint due to severe derailment damage in 1969, but the others were probably just given FYE rather than a full coat of blue. D7034 is seen on the scrap line at Old Oak Common in March 1972, looking fairly worn, although it had only been in service since January. (Grahame Wareham)

A standard Rail Blue 'Hymek', D7100, stands at Old Oak Common in April 1969. After the first few BFYE examples were done in 1966–67, the style altered slightly to what became the final BFYE version. On D7100, the full yellow ends extend more onto the cabsides than the early repaints (see the shot of D7058), and by now it has received the data panels designating it as a Class 35, which began to be applied to diesel and electric locos from the summer of 1968. 'Hymek' repaints continued until August 1971, with D7032 the last to go from green to blue. Withdrawals began the following month. (Grahame Wareham)

Rail Blue 7097 passes Gaer Junction, Newport, on 22 May 1971, with the 8A18 'Marshfield Milk', which was a daily feature at Cardiff. 7097's metal 'D' prefix has been painted over in blue, but not removed, which happened to many 'Hymeks'. Some locomotives had the 'D' removed altogether, others retained it to the end, and many had it painted over. A comprehensive list is difficult to give, as many had combinations of all these variations on different cabs! (Jim Sparks)

D7064 rolls through Port Talbot on 26 September 1971 in Rail Blue FYE. As on 7097, the 'D' prefix is painted over in blue. As stated earlier, this locomotive was formerly BSYP, complete with white window surrounds, and probably just had full yellow ends applied sometime after September 1969, as it retains the double arrow in the higher position, given when it was BSYP. (Dave Thorpe)

Above: The final 'Hymek' to get a repaint into Rail Blue from green was D7032 in August 1971. Withdrawals commenced one month later, and D7032 was claimed in May 1973. The loco is shown here, still in great external condition, at Old Oak Common on 4 September 1971. Few 'Hymeks' got more than one coat of blue, possibly only D7055 at Swindon, although a few did get depot repaints in 1973–74. (Trevor Casey)

Opposite above: 7076 stands at Cardiff Canton on 25 September 1971, showing the unusual removal of the cast alloy numbers on the driver's cabsides, and their replacement with standard BR style vinyls. The secondman's cabside on the far end retains the cast numbers though. (John Ireland)

Opposite below: D7038, previously seen ex works in 1967 with the shallower yellow ends, had acquired the standard version by the time it was photographed at Reading in May 1972, working a mixed freight. It has also had the data panels applied as normal, but retains the 'D' prefixes, these being neither removed nor painted over, and the higher placed arrow symbol. Withdrawal was just two months later. (Paul Townsend)

The 'Hymeks'

Above: Most of the 'Hymeks' had been withdrawn by the end of 1972, leaving just a few to soldier on for another two years. All survivors into 1973 and beyond were standard Rail Blue examples, such as D7001, here looking like it might not be long for this world at Twyford on 1 August 1973, with the 15.05 Paddington–Hereford. The worksplate has vanished and the 'D' prefix is painted over but remains in place. (Andy Kirkham)

Opposite above: Another late survivor was D7022, photographed at Westbury on 10 April 1974. Bits of its former green livery are starting to show through the blue over three years after it was repainted, and it is another with a painted out 'D' prefix. It had another 11 months to run in service before withdrawal in March 1975, two years after the 'Hymek Farewell' railtour was run! (Stephen Burdett)

Opposite below: The cabside detail on D7018, as seen at Old Oak Common on 8 February 1975, showing the cast alloy numbers. The chalked on 'NB' indicated that the steam-heat boiler had been isolated, which supposedly banned the loco from working passenger trains. However, this was blithely ignored by the WR! D7018 went into Rail Blue in February 1970. (Neil Phillips)

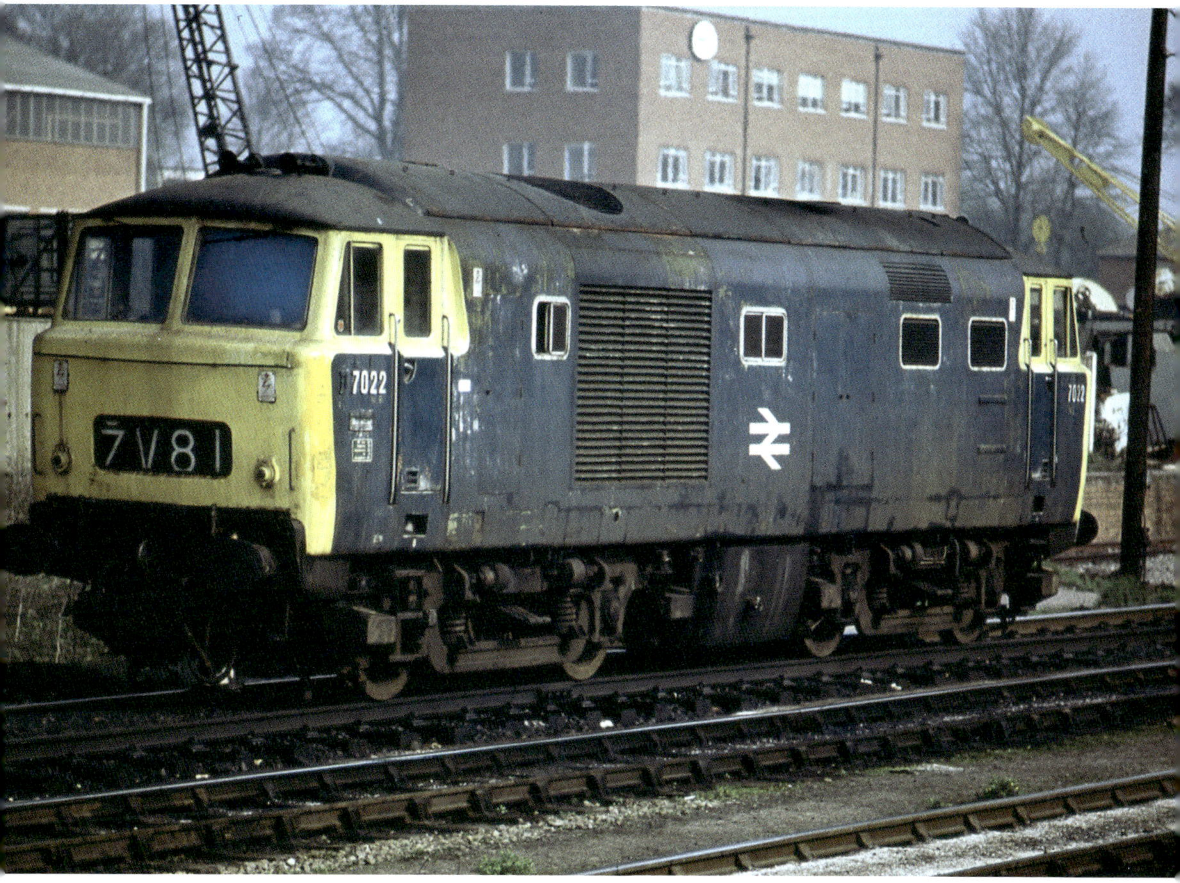
7V81
7022

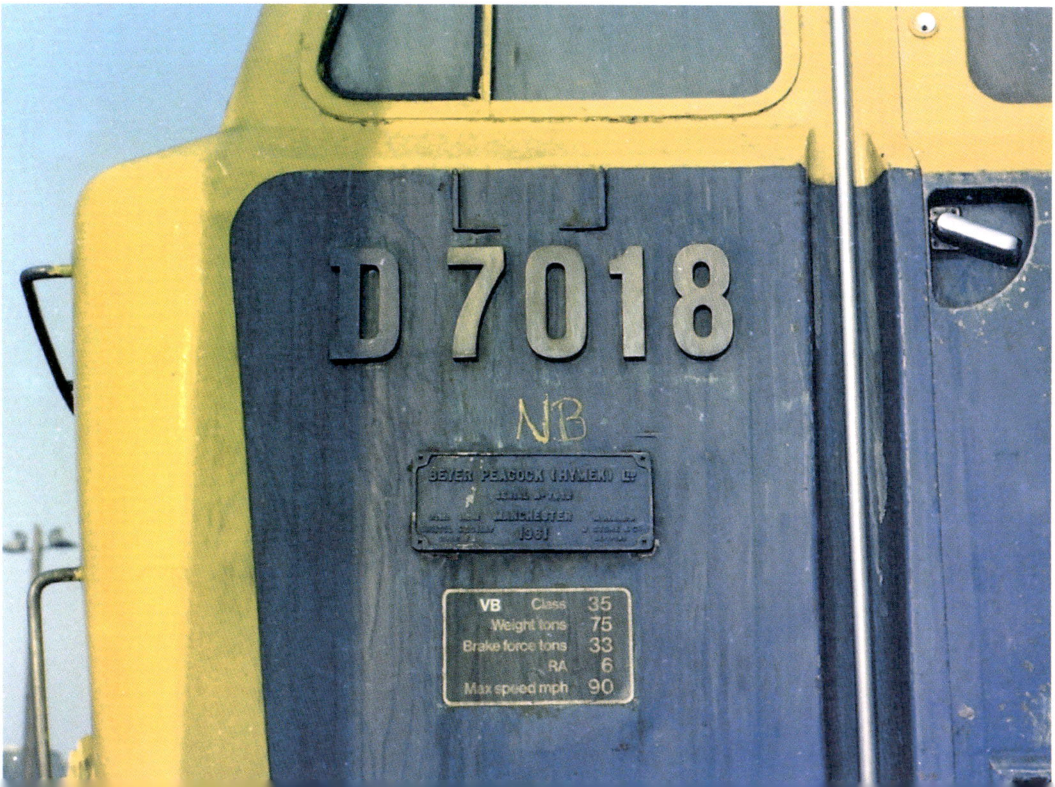
D7018
NB
BEYER PEACOCK (HYMEK) Ltd
MANCHESTER
1961
VB Class 35
Weight tons 75
Brake force tons 33
RA 6
Max speed mph 90

A contrast in the cabside of classmate D7017, at the same location on the same day. The alloy numbers have been removed as with 7076, and ostensibly the boiler is still in use. The data panel looks like it has been thrown together, as the numbers are unevenly placed. (Neil Phillips)

Chapter 6

Class 14

All of the WR hydraulics were withdrawn prematurely, and none worked for their designed lifespan, but the 56 D95xx 'Teddy Bears' had spectacularly short careers, on BR at least. Brand new D9540 stands in Swindon Works on 3 April 1965, in its shiny coat of Brunswick Green, with the paler cab and yellow bufferbeams. There was little variation from the as-delivered paint scheme, as they did not have the time, as, despite being built in 1964–65, all were withdrawn by April 1969. D9540 had just three years in traffic, and D9554 a mere two years and six months. None were ever given Rail Blue in BR ownership. (Jim Ireland)

D9500 stands at Cardiff Canton in May 1965 in pretty much new condition. It was one of the longest lived of the class, lasting five years until withdrawn in April 1969. The style of painted number used on these engines was unique. (Grahame Wareham)

D9502 is seen at work in the Forest of Dean in the autumn of 1965, still in as-delivered livery. Most were sold off to industry after their work dried up on BR and, happily, a third of the class survives to this day on heritage railways, including D9502, which is currently at the East Lancashire Railway. (Geoff Dowling)

Class 14

One minor livery variation on a D9500 was the addition of the shed allocation plate beneath the number on the cabsides. D9508 is seen at Cardiff Central alongside a Rail Blue 'Hymek' on a rake of blue/grey stock. Odd to think that the introduction of the D9500s paralleled that of blue/grey stock, D9500 entered service the same month the XP64 train was tried out, although all had gone long before all the stock was repainted into blue/grey. The view is undated, but was taken in 1967 or 1968, as that postdates the blue livery for a 'Hymek', and D9508 was withdrawn in October 1968. (Fred Castor)

Above: D9516 stands at Cardiff Canton on 21 November 1965. No shed allocation plate or data panel has been added yet, and although the D9500s just survived long enough to be allocated Class 14 under the 1968 scheme, as far as I am aware none ever had data panels applied. While it is possible that a couple may have done, I have certainly never come across a photo of one. (Jim Ireland)

Opposite above: Class 14s D9514 and D9518 lie withdrawn at Gloucester on 4 July 1969, presumably en route to their new owner, the National Coal Board (NCB) at Ashington. Both have shedplates showing they were last based at 86A Cardiff Canton, where they had been for the whole of their pitifully short careers, but apart from that, they remain in as-delivered condition. D9514 appears to lack the overhead live wires warning plates though. D9514 worked for another 16 years in industry after this shot was taken, being scrapped in 1985 by the NCB, but D9518 still exists, preserved on the West Somerset Railway. D9518 was the last Class 14 in normal revenue earning service on BR when switched off at Radyr on 19 April 1969, however a couple were resurrected for PW work around Hereford as late as May 1970! (Derek Jones)

Opposite below: Another dumped Class 14, D9527 stands at Gloucester on 4 July 1969. Pretty much in the as-delivered livery, it retains the 'D' prefix, as I believe they all did. Another one sold off to industry, its four years and three months in BR service were followed by 15 years in industrial use. Sadly, D9527 no longer exists, as it was scrapped in 1984 by the NCB at Ashington Colliery. (Derek Jones)

Class 14

FURTHER READING FROM

As Europe's leading transport publisher, we also produce a range of market-leading railway magazines and specials.

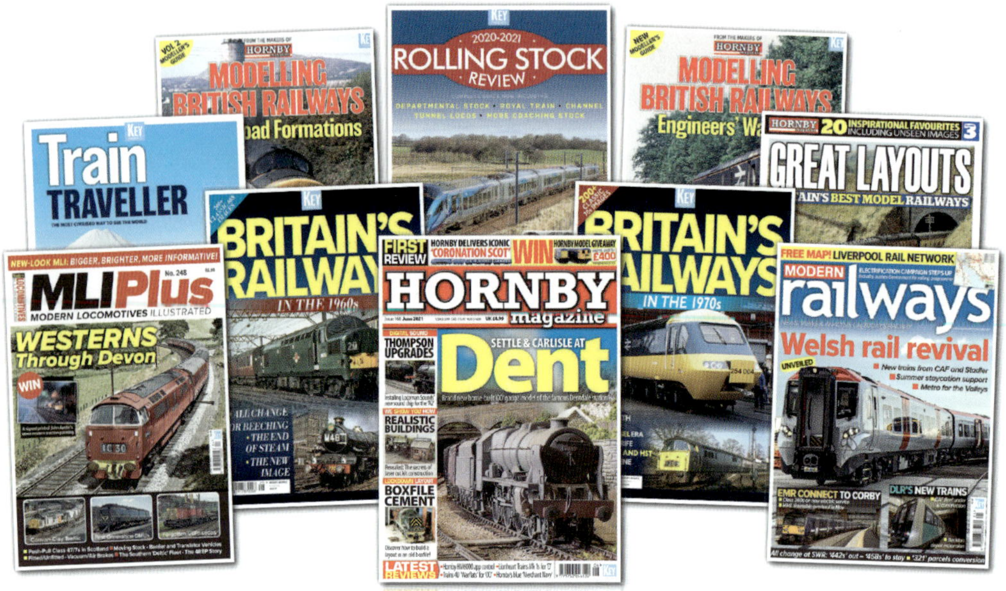

visit:
shop.keypublishing.com
for more details.

Your online home for **Modern railways**
www.keymodernrailways.com